GUIDE TO GROWING MARIJUANA OUTDOOR

Secret to Successful Outdoor Marijuana Cultivation

Ethel Iott

Copyright © [2023]

Table of Contents

Chapter 1: Introduction

Growing marijuana outdoors can be a rewarding and cost-effective way to cultivate your own cannabis plants. With the right knowledge and techniques, you can produce high-quality buds that are perfect for personal use. This guide will provide you with all the information you need to successfully grow marijuana outdoors, from choosing suitable strains to ensuring optimal growing conditions.

One of the advantages of outdoor cultivation is that it allows the marijuana plants to benefit from natural sunlight and fresh air, resulting in healthy and robust growth. Furthermore, growing marijuana outdoors can be more budget-friendly compared to indoor growing, as you can avoid the cost of artificial lighting and climate control equipment.

In this guide, we will cover important aspects such as selecting the right strains that are well-suited for outdoor cultivation, preparing the soil for planting, providing adequate nutrients and water, and protecting your plants from pests and diseases. We will also provide tips on when to harvest your plants and how to properly dry and cure the buds for optimal flavor and potency.

Whether you are a beginner or an experienced grower, this guide will serve as a comprehensive resource to help you successfully grow marijuana outdoors and enjoy the fruits of your labor.

Overview of Outdoor Marijuana Growing

Outdoor marijuana growing refers to cultivating marijuana plants in an open-air environment, such as a garden or a field, as opposed to growing indoors or in a greenhouse. This method of growing cannabis has been practiced for centuries and has gained popularity due to its numerous advantages.

One of the main benefits of outdoor marijuana growing is the lower cost associated with it. Unlike indoor growing, which requires the use of expensive equipment, outdoor growing utilizes natural sunlight, reducing the need for artificial lighting. This saves money on electricity bills and equipment costs, making it a more cost-effective option for many growers.

Additionally, outdoor growing allows cannabis plants to benefit from the natural nutrients present in the soil. This often results in larger, healthier plants with higher yields compared to those grown indoors. Outdoor cultivation also provides more space for the plants to grow, allowing them to reach their full potential.

Another advantage of outdoor marijuana growing is the ability to harness the power of the sun. Sunlight provides a full spectrum of light that is essential for a plant's growth, development, and production of cannabinoids. This natural light source also provides UV rays, which can enhance the potency of the plant's resin.

However, outdoor growing also has its challenges. One of the biggest challenges is the lack of control over environmental factors. Growers have to contend with weather changes, pests, and potential theft or vandalism. To mitigate these risks, growers often employ various techniques such as installing fences, using organic pest control methods, and monitoring weather patterns.

Overall, outdoor marijuana growing is a popular choice for growers who want to take advantage of natural sunlight, reduce costs, and produce larger yields. It allows for the cultivation of high-quality cannabis plants with potent resin profiles. However, it does require careful planning and attention to detail to ensure a successful and fruitful harvest.

Benefits of Growing Marijuana Outdoors

Growing marijuana outdoors has several benefits that make it an attractive option for many growers. One of the main advantages is the cost savings. Growing cannabis indoors can be expensive, as it requires specialized equipment such as lights, fans, and ventilation systems, as well as the cost of electricity to power them. By growing marijuana outdoors, growers can take advantage of natural sunlight, reducing the need for artificial lighting and saving on electricity costs.

Another benefit of growing marijuana outdoors is the potential for larger yields. Outdoor plants have the opportunity to grow taller and wider than those grown indoors, allowing for greater bud production. Additionally, outdoor plants have the advantage of a longer vegetative phase due to the longer daylight hours during the summer months. This extended growth period can lead to larger plants and ultimately, larger yields.

Outdoor growing also allows for a more organic cultivation process. Without the need for artificial lighting or climate control, growers can rely on natural elements to cultivate their plants. This can result in a more sustainable and environmentally-friendly growing process.

Finally, growing marijuana outdoors can also be a less stressful experience for the grower. Indoor growing requires continuous monitoring and management to maintain the proper environmental conditions for the plants. Outdoor plants, on the other hand, are better able to adapt to changes in weather and are generally less prone to pests and diseases. This means less time and effort spent on maintenance and more time to enjoy the process of growing.

Overall, growing marijuana outdoors can offer cost savings, larger yields, a more organic cultivation process, and a less stressful experience for the grower. It allows for the use of natural sunlight, which can lead to larger plants and ultimately, bigger yields. It also allows for a more sustainable and environmentally-friendly growing process, as growers can rely on natural elements rather than artificial lighting and climate control. With less monitoring and maintenance required, growing marijuana outdoors can be a more enjoyable and hassle-free experience.

Legal Considerations and Restrictions

When it comes to growing marijuana outdoors, there are several legal considerations and restrictions that must be taken into account. These laws may vary from state to state and even within different municipalities. It is essential for growers to be familiar with and abide by all local laws and regulations to avoid legal consequences.

One of the most important legal considerations is whether or not marijuana cultivation is legal in the area. In some states, marijuana cultivation is legal for both medical and recreational use, while in others it may only be legal for medical use or strictly prohibited altogether. It is vital for growers to understand the specific laws in their area to ensure compliance.

There may also be restrictions on where marijuana can be cultivated outdoors. Some states prohibit outdoor cultivation altogether and require it to be grown indoors or in specific locations. Other areas may have zoning restrictions that limit where marijuana can be grown, such as a certain distance from schools or parks.

Additionally, growers may need to obtain a license or permit to legally grow marijuana outdoors. These licenses typically require meeting certain criteria, such as security measures, record-keeping, and inspections. Failure to obtain the necessary permits can result in fines or even criminal charges.

Finally, it is important to consider the impact of federal law. While some states have legalized marijuana cultivation, it remains illegal under federal law. This means that even if growers are in compliance with state laws, they can still face legal consequences on a federal level.

Overall, growing marijuana outdoors requires careful consideration of local laws and regulations. Legal restrictions can vary greatly, so it is crucial for growers to do their due diligence and research the specific requirements in their area. This will help ensure that they are compliant with the law and can cultivate marijuana in a legal and safe manner.

Climate and Location Requirements

When it comes to growing marijuana outdoors, two important factors to consider are climate and location requirements. These play a crucial role in determining the success and overall health of the plants.

Climate: Marijuana plants thrive in a specific temperature and humidity range. They prefer a warm climate with temperatures between 70-85 degrees Fahrenheit during the day and slightly cooler temperatures at night. The ideal humidity level for marijuana plants ranges between 40-60%. Hot and dry climates can cause stress to the plants, while cold and wet climates can lead to mold and other plant diseases. Therefore, regions with a Mediterranean-like climate or similar conditions are often ideal for outdoor marijuana cultivation.

Location: The location of the outdoor marijuana garden is equally important as the climate. It should be chosen wisely to maximize sunlight exposure, airflow, and privacy. Marijuana plants require at least 6-8 hours of direct sunlight per day to grow and produce high-quality buds. Therefore, selecting a location that is not shaded by tall trees or buildings is crucial. Additionally, proper airflow is essential to prevent the development of mold and mildew. Thus, an open area or a spot with good wind circulation is preferred. Lastly, privacy is necessary to avoid any unwanted attention or theft. Concealing the garden from public view can be achieved through the use of privacy fences or strategic positioning of the plants.

Overall, when growing marijuana outdoors, it is important to choose a location that provides the right climate and environmental conditions. This will ensure that the plants can grow and flourish, yielding a bountiful harvest of high-quality buds. Conducting thorough research and consulting with experienced growers in the area can also provide valuable insights and guidance on the best practices for outdoor marijuana cultivation.

Chapter 2: Preparing for Outdoor Marijuana Growing

When preparing for outdoor marijuana growing, there are several important factors to consider. First, it is crucial to choose the right location. Marijuana plants need a lot of sunlight, so selecting a spot that receives direct sunlight for at least 6-8 hours a day is essential. It is also important to choose a spot that is secluded and hidden from view to ensure privacy and security.

Next, it is important to prepare the soil properly. Marijuana plants require well-drained soil that is rich in organic matter. Before planting, it is recommended to mix in compost, worm castings, and other organic fertilizers to improve the soil quality.

Additionally, it is important to consider the watering requirements of outdoor marijuana plants. These plants require a lot of water, especially during the hot summer months. Providing a consistent watering schedule and ensuring proper drainage is essential to prevent overwatering and root rot.

Lastly, it is important to consider pest control measures. Outdoor marijuana plants are susceptible to pests such as aphids, spider mites, and caterpillars. Implementing organic pest control methods, such as introducing beneficial insects or using neem oil, can help protect the plants from these pests. By taking these factors into consideration and properly preparing for outdoor marijuana growing, growers can increase their chances of successfully cultivating healthy and productive marijuana plants.

Selecting the Right Strain

When it comes to growing marijuana outdoors, selecting the right strain is crucial for a successful and bountiful harvest. Different strains have different characteristics and growing requirements, so it's important to choose a strain that will thrive in your specific climate and conditions.

One of the first factors to consider when selecting a strain for outdoor growing is the climate in which you live. Some strains are more resilient to cold temperatures, while others prefer a warmer climate. Understanding your local climate and selecting a strain that is well-suited to it will increase the chances of a successful grow.

Another important consideration is the size of the plants. Outdoor growers typically have more space to work with, so it's important to choose a strain that will fit well in your intended grow space. Some strains can grow quite tall and bushy, while others stay more compact and shorter. Think about how much space you have available and choose a strain that will be able to grow and flourish in that space.

Finally, consider the desired effects and potency of the strain. Different strains have different levels of THC and CBD, so think about what kind of experience you're looking for. If you're looking for a strain with a high THC content and a strong psychoactive effect, choose a strain that is known for its potency. On the other hand, if you're looking for a strain with a lower THC content and more CBD for medicinal purposes, select a strain that is known for its therapeutic properties.

In addition to these factors, it's also important to research the specific growing requirements of each strain you are considering. Some strains may require more attention and care, while others are more resilient and can handle different types of conditions. Make sure that you are prepared to meet the specific needs of the strain you choose.

Overall, selecting the right strain for growing marijuana outdoors requires careful consideration of your local climate, available space, desired effects, and the specific needs of the strain. By taking these factors into account, you can increase your chances of a successful outdoor grow and a bountiful harvest of high-quality marijuana.

Germination and Seedling Care

Germination and seedling care are crucial steps in growing marijuana outdoor. Ensuring the proper conditions for germination will result in healthy seedlings and a successful growing season.

To germinate marijuana seeds, start by placing the seeds in a glass of water or in a damp paper towel. Keep the seeds in a warm and dark place, such as a closet or cabinet, for 24-48 hours. This will allow the seeds to absorb water and activate the germination process.

Once the seeds have sprouted roots, carefully transfer them to a small pot with a well-draining soil mix. Place the pot in a sunny location where the seedlings can receive at least six hours of direct sunlight per day. If sunlight is limited, use fluorescent lights or LED grow lights to supplement the natural light.

Seedlings have delicate root systems, so it's important to avoid overwatering. Water the seedlings only when the top inch of the soil feels dry. Use a gentle misting spray bottle to water the seedlings to prevent overwatering or disturbing the roots.

Additionally, it is important to protect seedlings from harsh weather conditions, such as strong winds or extreme temperatures. Consider using a small greenhouse or cold frame to provide a protected environment for the seedlings.

Regularly monitor the seedlings for any signs of pests or disease. Common pests for outdoor marijuana grows include aphids, spider mites, and caterpillars. If pests are present, use organic pest control methods or consult with a professional for guidance.

As the seedlings grow, they will need to be gradually acclimated to outdoor conditions. This process, known as hardening off, involves slowly introducing the seedlings to direct sunlight, wind, and lower humidity levels. Start by placing the seedlings outdoors for a few hours each day and gradually increase the amount of time they spend outside over the course of a week. During the seedling stage, it is important to provide appropriate nutrients to support healthy growth. Use a balanced, organic fertilizer formulated for cannabis plants.

By following these steps for germination and seedling care, you can ensure a strong start for your marijuana plants and increase your chances of a successful outdoor grow season. Proper care during these early stages will set the foundation for healthy, productive plants later on.

Soil Preparation and Amendments

Soil preparation and amendments play a crucial role in the successful cultivation of marijuana outdoors. It is important to start with high-quality soil that is rich in essential nutrients and has good drainage. This can be achieved by mixing in compost or other organic matter to improve the soil's fertility and structure.

Before planting, it is also recommended to have the soil tested to determine its pH level. Marijuana plants prefer slightly acidic soil, with a pH between 6.0 and 6.8. If the soil is too acidic or alkaline, amendments like lime or sulfur can be added to adjust the pH to the optimal range.

In addition to pH adjustments, it is important to provide the right balance of essential nutrients for the marijuana plants. Common soil amendments used for this purpose include bone meal for phosphorus, blood meal or fish emulsion for nitrogen, and kelp meal for trace minerals. These amendments can be added to the soil during the preparation process or applied as top dressings throughout the growing season.

Proper soil preparation and amendments can also help improve the soil's water retention capacity. This is especially important for marijuana plants grown outdoors, as they rely on natural rainfall for hydration. Adding organic matter like compost or coconut coir can help retain moisture in the soil and reduce the need for frequent watering.

It is important to note that marijuana plants have different nutrient requirements at different stages of growth. During the vegetative stage, they require higher levels of nitrogen to promote leafy growth. As the plants transition to the flowering stage, a switch to a nutrient blend higher in phosphorus and potassium is necessary to support the development of buds. Regular soil testing and adjustments can help ensure that the plants receive the right nutrients at each stage.

Proper soil preparation and amendments not only provide the necessary nutrients for marijuana plants but also create an environment that promotes healthy root development. This, in turn, leads to strong and vigorous plants that are more resistant to pests and diseases.

In conclusion, soil preparation and amendments are essential for successful outdoor marijuana cultivation. Starting with high-quality soil, adjusting the pH, and providing the right balance of nutrients are all crucial factors in creating a favorable growing environment. By taking these steps, growers can give their marijuana plants the best chance at thriving and producing high-quality buds.

Choosing the Right Containers or Beds

When growing marijuana outdoors, choosing the right containers or beds is crucial for the health and productivity of your plants. The container or bed you choose will determine the root growth, drainage, and overall health of your plants.

When it comes to containers, it is important to choose ones that are large enough to accommodate the root system of your plants. A general rule of thumb is to use a container that is at least 5 gallons in size, but larger containers, such as 10 or even 20 gallons, are recommended for bigger plants. The larger the container, the more room the roots have to grow, which in turn leads to healthier and more productive plants.

In terms of materials, fabric containers are becoming increasingly popular among marijuana growers. Fabric containers allow air to flow through the root system, promoting healthy root growth and preventing root-binding. They also have excellent drainage capabilities, preventing water from pooling at the bottom of the container and potentially causing root rot.

If you prefer to grow your marijuana plants directly in the ground, raised beds are a great option. Raised beds provide a deeper and more fertile soil profile, which encourages healthy root growth and maximizes the nutrient uptake of your plants. Additionally, raised beds provide better drainage compared to normal garden beds, ensuring that excess moisture doesn't cause waterlogged roots.

When choosing the right containers or beds for growing marijuana outdoors, it is also important to consider the portability and accessibility of the containers. If you anticipate needing to move your plants for any reason, choosing containers with handles or wheels can make the process much easier. Likewise, if you have limited mobility, raised beds at a comfortable height can provide easier access for tending to your plants.

Lastly, consider the aesthetics of your garden. If you want your marijuana plants to blend in or if you have limited space, you may opt for more discreet containers or beds that can be easily hidden or arranged in a small space. On the other hand, if you want your plants to be a focal point in your outdoor space, larger and more visually appealing containers or beds may be the way to go.

Overall, choosing the right containers or beds for growing marijuana outdoors is essential for the success of your plants. Consider the size, material, drainage capabilities, portability, accessibility, and aesthetics when making your decision. By providing your plants with the right environment, they will have the best chance of thriving and producing high-quality buds.

Implementing Security and Stealth Measures

Implementing security and stealth measures for growing marijuana outdoors is crucial to ensure the safety and success of the operation. Growing marijuana outdoors can be risky due to its illegal status in many areas, so taking steps to protect the crop is essential.

One of the first steps in implementing security measures is choosing a remote location for the grow site. The farther away the site is from populated areas, the lower the risk of discovery. It's also important to select a site with natural barriers, such as dense vegetation or rock formations, to help conceal the plants from view.

Another important security measure is installing a perimeter fence around the grow site. This will help prevent unauthorized access and deter potential thieves. The fence should be tall and sturdy, with no gaps or weak points.

In addition to physical security measures, it's important to implement stealth measures to avoid drawing attention to the grow site. This includes using camouflage netting or tarps to conceal the plants, as well as choosing strains that blend in with the surrounding vegetation. It's also important to avoid any telltale signs, such as strong odors or excessive noise, that could alert others to the presence of the grow site.

Finally, it's important to regularly monitor the grow site for any signs of intrusion or theft. This can be done through the use of cameras or motion sensors, which can be set up around the perimeter of the grow site. Regular visits to the site to check on the plants and ensure their well-being can also help identify any potential security breaches.

Overall, implementing security and stealth measures for growing marijuana outdoors is a necessary step to protect the crop from theft or discovery. By selecting a remote location, installing a perimeter fence, using camouflage and stealth techniques, and regularly monitoring the grow site, growers can minimize the risks associated with outdoor cultivation and increase the chances of a successful harvest.

Chapter 3: Outdoor Marijuana Growing Techniques

Outdoor marijuana growing techniques involve transplanting and watering the plants to ensure optimum growth and yield. Transplanting is the process of moving the plants from one container to a larger one or directly into the ground. This helps the roots to spread out and gain access to more nutrients and water. Transplanting should be done carefully to avoid damaging the delicate roots.

Watering is a crucial aspect of outdoor marijuana cultivation. The plants need a consistent supply of water to grow and thrive. The frequency and amount of water required may vary depending on the climate and the stage of growth. It is important to water the plants deeply and at the base to ensure that the roots absorb the water efficiently.

Monitoring soil moisture levels is essential to prevent over or underwatering. Overwatering can lead to root rot and other problems, while underwatering can stress the plants and result in stunted growth. It is recommended to water the plants in the morning or evening to minimize evaporation.

Implementing these outdoor marijuana growing techniques, such as transplanting and proper watering, can help growers achieve healthy and productive plants. However, it is important to note that each growing environment is unique, and growers should adapt their techniques to suit their specific conditions.

Transplanting Marijuana Seedlings

Transplanting marijuana seedlings is an essential step in the process of growing marijuana outdoors. It is important to transplant seedlings at the right time to ensure their proper growth and development.

The ideal time to transplant marijuana seedlings is when they have developed a strong root system and are about 2-3 inches tall. It is crucial to transplant them before they become root-bound in their initial containers. Transplanting at this stage will help the plants adapt better to the outdoor environment and maximize their growth potential.

To transplant marijuana seedlings, start by selecting a suitable outdoor location with adequate sunlight and well-drained soil. Dig a hole that is twice as wide and deep as the seedling's original container. Carefully remove the seedling from its container, making sure to handle the roots gently to avoid damage.

Place the marijuana seedling in the hole and backfill it with soil, ensuring that the plant is at the same level it was in its original container. Firmly press down the soil around the seedling to eliminate any air pockets. Water the plant thoroughly after transplanting to help settle the soil and provide hydration to the roots.

It is also advisable to provide some form of protection for the newly transplanted seedlings, such as a temporary shelter or a layer of mulch, to shield them from harsh weather conditions, pests, and other potential threats.

After transplanting, it is important to monitor the seedlings closely and provide them with the necessary care and nutrients for optimal growth. Additionally, keep an eye out for any signs of stress or nutrient deficiencies, and address them promptly.

Transplanting marijuana seedlings for outdoor growing is a crucial step in ensuring successful cultivation. It helps the plants adapt to their new environment and provides them with the space and nutrients they need to thrive. By following the proper transplanting techniques and providing ongoing care, outdoor growers can set their marijuana plants up for a successful growing season and a bountiful harvest.

Watering and Feeding Marijuana Plants

When it comes to growing marijuana plants outdoors, watering and feeding are two crucial aspects to consider. Proper watering is essential for healthy plant growth, while feeding ensures that the plants receive all the necessary nutrients they need to thrive.

Watering marijuana plants outdoors can be a bit trickier compared to indoor grows. The amount of water needed depends on various factors such as the weather conditions, soil type, and the size of the plants. Generally, it is recommended to water the plants deeply and less frequently rather than shallowly and frequently. This encourages the plants to establish deeper root systems, making them more resistant to drought conditions.

In terms of nutrients, marijuana plants require a balanced diet of macronutrients (nitrogen, phosphorus, and potassium) as well as micronutrients (iron, magnesium, and calcium). These nutrients are typically provided through fertilizers, either organic or synthetic. It is important to choose a fertilizer that is specifically formulated for marijuana plants, as they have specific nutrient requirements.

Feeding marijuana plants should be done in accordance with the plant's growth stage. During the vegetative stage, plants require needs high levels of nitrogen to enable leaf and stem growth. During the flowering stage, the plants require higher levels of phosphorus and potassium to encourage bud development. Always follow the instructions on the fertilizer label and adjust the feeding schedule accordingly.

Another important aspect of feeding marijuana plants is pH levels. The optimal pH range for marijuana plants is between 6.0 and 6.5. This ensures that the plants can properly absorb nutrients from the soil. It is recommended to regularly test the pH levels of the soil and make any necessary adjustments using pH up or pH down products.

In addition to regular watering and feeding, it is important to also consider other factors that can affect the overall health of the plants. These include proper drainage, adequate sunlight, and protection from pests and diseases. It is also crucial to monitor the plants closely for any signs of nutrient deficiencies or excesses, as well as any potential pest or disease issues.

Overall, providing proper watering and feeding for marijuana plants grown outdoors is essential for healthy and productive growth. It is important to keep in mind the specific needs of the plants at different stages of growth and adjust the watering and feeding accordingly. By providing the right nutrients and maintaining optimal conditions, growers can ensure that their marijuana plants thrive and produce high-quality yields.

Training and Pruning Techniques

Training and pruning techniques are crucial for growing marijuana outdoors. These techniques help ensure that the plants receive adequate light and air circulation, resulting in healthier and more productive plants.

One commonly used training technique is the low-stress training (LST) method. This method involves gently bending the branches of the plant to create an even and open canopy. By doing so, more light can reach the lower parts of the plant, ensuring better bud development. LST can be done by using soft tying materials to gently pull down branches and secure them to stakes or other supports.

Another important technique is topping. Topping involves cutting off the main stem of the plant, which encourages the growth of lateral branches. This results in a bushier plant with multiple colas, increasing the overall yield. Topping should be done during the vegetative stage when the plant has at least five to six nodes.

Pruning is also essential for outdoor cultivation. Removing any yellow, dead, or damaged leaves helps to prevent the spread of diseases and improves airflow around the plant. Cleaning up the lower branches and leaves can also reduce the risk of pests and mold.

Overall, training and pruning techniques are vital for outdoor marijuana cultivation. By implementing these techniques, growers can ensure proper light distribution, enhance bud development, increase yields , and maintain healthier plants. It is important to note that these techniques should be done with caution and precision, as improper pruning or training can cause stress to the plant or lead to reduced yields. It is recommended to research and learn about the specific needs and characteristics of the marijuana strain being grown, as different strains may require different training and pruning techniques. Additionally, regular monitoring and observation of the plants' growth and development is essential to make necessary adjustments and ensure successful cultivation.

Pest and Disease Control Methods

When it comes to growing marijuana outdoors, the threat of pests and diseases is a constant concern. However, there are several effective methods for controlling and preventing these issues.

One of the most common pests that affect outdoor marijuana plants is aphids. These small insects suck the sap from the leaves, causing wilting and discoloration. To control aphids, you can use a commercial insecticide or create a homemade solution using neem oil and water. Neem oil has insecticidal properties and is safe to use on marijuana plants. Another common pest is the spider mite, which creates webs on the leaves and sucks the plant's sap. Spider mites can be controlled using insecticidal soap or by spraying the plants with a mixture of water and dish soap.

For diseases, the most common one affecting outdoor marijuana plants is powdery mildew. This fungal infection forms a white powdery coating on the leaves, hindering photosynthesis. To prevent powdery mildew, it is essential to provide proper air circulation by spacing plants adequately and removing any dead or infected leaves. Spraying the plants with a mixture of water, baking soda, and dish soap can also help control powdery mildew.

To avoid attracting pests and diseases in the first place, it is crucial to practice good hygiene and sanitation in your garden. This includes regularly removing any dead or decaying plant material, as it can attract pests and harbor disease spores.

Furthermore, it is essential to choose disease-resistant strains of marijuana when possible. These strains have been bred specifically to be less susceptible to common diseases, giving your plants a better chance of staying healthy. Additionally, planting companion plants that naturally repel pests can be beneficial. For example, marigolds and lavender can deter aphids, while garlic can repel spider mites.

Last but not least, early detection and prevention of pests and illnesses depend on routinely checking your plants for symptoms. Early detection decreases the chance of the issue spreading and enables speedier response.

In conclusion, maintaining suitable pest and disease management techniques is crucial for a productive and healthy crop when growing marijuana outdoors. You can maintain your plants healthy and flourishing throughout the growing season by combining preventative measures, good cleanliness, and prompt intervention.

Managing Environmental Factors (light, temperature, humidity)

Managing environmental factors is crucial for successfully growing marijuana outdoors. Light, temperature, and humidity all play significant roles in the growth and development of marijuana plants.

Light is perhaps the most important environmental factor to consider when growing marijuana outdoors. Marijuana plants require a sufficient amount of light to carry out photosynthesis and produce energy. Ideally, they need at least 12 hours of direct sunlight per day. However, excessive heat and intense sunlight can damage the plants, so it is important to provide some shade during the hottest part of the day.

Temperature also plays a critical role in the growth of marijuana plants. Generally, marijuana plants thrive in temperatures between 70°F and 85°F during the day and around 60°F to 70°F at night. Extreme temperature fluctuations can stress the plants and affect their overall health and productivity. It is important to monitor the temperature and provide adequate ventilation to prevent overheating or cold drafts.

Humidity levels should also be carefully managed when growing marijuana outdoors. High humidity can lead to the growth of mold, mildew, and other fungi, which can be detrimental to the plants. However, too low humidity can cause the plants to dry out and wilt. It is important to maintain a humidity level between 40% and 60% during the daytime and slightly lower at night. This can be achieved by providing adequate air circulation and using dehumidifiers or humidifiers if necessary.

To effectively manage these environmental factors, it is important to regularly monitor and adjust as needed. This can be done by using temperature and humidity gauges, as well as observing the plants for any signs of stress or damage. If the temperature is too high, shade cloths or water misters can be used to cool down the area. If the humidity is too high, fans or dehumidifiers can be used to circulate the air and reduce moisture.

In addition to managing these factors, it is also important to consider the specific needs of different marijuana strains. Some strains may require more or less light, temperature, or humidity than others. It is important to research and understand the specific requirements of the strains being grown to ensure optimal growth and yield.

By effectively managing environmental factors such as light, temperature, and humidity, growers can create an ideal growing environment for marijuana plants. This can lead to healthier plants, higher yields, and better quality buds. However, it requires careful monitoring and adjustments to ensure that the plants are receiving the optimal conditions for growth.

Chapter 4: Harvesting and Beyond

Harvesting and Beyond is an essential guide for anyone interested in growing marijuana outdoors. This comprehensive chapter takes you through all the steps necessary to successfully cultivate cannabis plants in an outdoor environment. The chapter provides detailed information on selecting the right strain, preparing the soil, understanding the climate, and preventing pests and diseases. It also covers the important topic of harvesting, providing tips on when and how to harvest your plants for maximum yield and potency. Furthermore, the chapter goes beyond harvesting and explores the various post-harvest processes such as curing, drying, and storing marijuana. Harvesting and Beyond is a valuable resource for both novice and experienced marijuana growers looking to optimize their outdoor cultivation.

When and How to Harvest Outdoor Marijuana

Outdoor marijuana plants are typically ready to be harvested in the late summer or early fall, depending on the specific strain and growing conditions. The ideal time to harvest is when the trichomes, or resin glands, on the buds have turned from clear to a milky white color. This indicates that the THC, the psychoactive compound in marijuana, is at its peak potency. Some strains may also develop amber-colored trichomes, which can indicate a more sedative effect.

To harvest outdoor marijuana, it is advised to wait until the plants are dry and there has been no rain for a few days. This will help prevent mold and mildew from developing on the buds. When harvesting, it is best to use clean, sharp scissors or shears to carefully cut the buds from the plant. It is important to handle the buds as gently as possible to prevent damage or trichome loss.

After the buds are harvested, they should be hung upside down in a dark, well-ventilated area with low humidity to dry. This process, known as curing, can take anywhere from a few days to a few weeks, depending on the desired moisture content. It is important to monitor the buds regularly and make sure they are not drying too quickly or becoming too dry, as this can affect the quality of the final product.

Once the buds are completely dry, they can be trimmed of any excess leaves and stems. Some growers prefer to leave a small amount of sugar leaves on the buds, as they can contribute to the overall flavor and potency. The trimmed buds should then be stored in airtight containers, such as glass jars, in a cool, dark place.

It is important to note that outdoor marijuana plants can be susceptible to pests and diseases, so it is important to regularly inspect the plants for any signs of damage or infestation. If any issues are detected, it may be necessary to harvest earlier than anticipated to prevent further damage to the crop.

Overall, knowing when and how to harvest outdoor marijuana is crucial in order to obtain the highest quality buds. Proper timing and drying techniques are key to preserving the potency and flavor of the final product. By following these guidelines, outdoor growers can ensure a successful and rewarding harvest.

Proper Drying and Curing Techniques

Proper drying and curing techniques are crucial for ensuring the highest quality marijuana when growing outdoor. Drying and curing are two distinct steps in the process of harvesting marijuana, and both are important for maximizing the potency, flavor, and overall smoking experience.

Drying is the first step after harvesting your marijuana plants. It should be done in a controlled environment with low humidity and good air circulation. The plants should be hung upside down, allowing air to circulate around them and dry them slowly. It is important to avoid drying the plants too quickly, as this can result in a harsh and unpleasant smoke. The drying process usually takes around 7-10 days, but it may vary depending on the climate and humidity levels in your area.

After the plants are dry, they should be trimmed and placed in jars for the curing process. Curing involves placing the buds in airtight containers, such as glass jars, and allowing them to rest for several weeks. During this time, moisture is gradually released from the buds, resulting in a smoother smoke and enhanced flavor. The jars should be opened daily for a short period of time to allow fresh air to circulate and prevent mold from forming.

Proper drying and curing techniques are essential for achieving the desired potency, flavor, and aroma of the marijuana. Rushing the drying process can lead to a harsh smoke that lacks the full flavor and potency potential. Similarly, skipping or rushing the curing process can result in a less smooth smoke and a loss of some of the terpenes that give the marijuana its unique aroma and flavor profile.

During the drying process, it is important to keep an eye on the humidity levels in the drying area. Too high humidity can lead to mold growth, while too low humidity can cause the plants to dry too quickly. Maintaining a humidity level between 45-55% is ideal for drying marijuana.

When it comes to curing, the length of time will depend on personal preference, but a minimum of two weeks is recommended. Opening the jars daily and checking for any signs of mold or excess moisture is crucial during this time. If mold is detected, it is important to remove the affected buds immediately to prevent it from spreading.

Proper drying and curing techniques are crucial steps in the marijuana cultivation process. They ensure that the final product is of the highest quality, with maximum potency, clean burn, and a rich flavor profile. Taking the time to properly dry and cure your harvested marijuana will result in a superior smoking experience and a product that you can be proud of.

Storing and Preserving Marijuana Buds

When it comes to storing and preserving marijuana buds, there are several important factors to consider. These include humidity, temperature, light exposure, and air circulation. Proper storage will help maintain the potency, flavor, and overall quality of the buds, ensuring a pleasant and effective experience.

One key element to preserve marijuana buds is controlling the humidity levels. Ideally, buds should be stored in airtight containers with a humidity range of 55-62%. This prevents the buds from becoming too dry or too moist, which can degrade their quality. Using humidity packs or boveda packs can help regulate the humidity levels within the containers.

Temperature is another crucial factor. It is recommended to store marijuana buds in a cool and dark location, as excessive heat can adversely affect their potency and flavor. The optimal temperature range for storage is between 60-70°F (15-21°C).

Light exposure can also impact the quality of marijuana buds. Exposure to ultraviolet (UV) light can degrade the cannabinoids and terpenes, diminishing the potency and flavor. Therefore, it is recommended to store buds in opaque containers or in a dark place to minimize light exposure.

Lastly, air circulation should be considered when storing marijuana buds. Proper air circulation helps prevent the growth of mold and mildew, which can ruin the buds. Storing them in airtight containers with minimal air exchange is key to preserving their freshness and quality.

In addition to these factors, it is also advisable to handle the buds with clean hands or utensils to avoid contamination. The use of gloves can be helpful in maintaining cleanliness. It is also important to avoid touching the buds unnecessarily to prevent the loss of trichomes, which contain many of the cannabinoids and terpenes responsible for the desired effects and flavors.

When it comes to long-term storage, vacuum-sealed bags or jars can be used to minimize exposure to air and moisture. These methods can help extend the shelf life of the buds and maintain their potency and quality for a longer period.

Ultimately, proper storage and preservation of marijuana buds are vital to ensure a pleasant and effective experience. By controlling humidity, temperature, light exposure, and air circulation, users can maximize the freshness, potency, and flavor of their buds, allowing for a more enjoyable and satisfying consumption.

Understanding and Utilizing Trim and Byproducts

Understanding and utilizing trim and byproducts is essential for growing marijuana outdoors. When trimming the marijuana plants, growers are left with a significant amount of excess foliage, stems, and bud trimmings. Many novice growers may overlook the potential value of these byproducts, but they can actually be utilized in various ways.

Firstly, trim and byproducts can be used to make potent concentrates and extracts. The trimmings contain valuable cannabinoids and terpenes that can be extracted using solvents like butane or CO_2. These extracts can then be used to create various marijuana products like oils, edibles, and tinctures.

Additionally, trim and byproducts can be used to create compost. Composting is a natural and eco-friendly way to dispose of organic waste. By composting the trimmings, growers can create nutrient-rich soil that can be used to grow healthier marijuana plants in the future.

Furthermore, some growers may choose to sell or trade their trim and byproducts to other marijuana enthusiasts. With the growing market for marijuana products, there is a demand for various byproducts like fan leaves, stems, and even non-smokable buds. This can be a way for growers to make an extra income or build relationships within the marijuana community.

In conclusion, understanding and utilizing trim and byproducts is crucial for maximizing the potential of outdoor marijuana cultivation. By extracting cannabinoids and terpenes from trimmings, creating compost, or selling/trading byproducts, growers can make the most of their harvest and contribute to a more sustainable and profitable cultivation process.

Planning for Future Outdoor Marijuana Grows

When it comes to planning for future outdoor marijuana grows, there are several factors that need to be considered. First and foremost, choosing the right location is crucial. The ideal location should have a favorable climate, plenty of sunlight, and a secure perimeter to prevent theft or unauthorized access.

Next, it is important to consider the size of the operation and the amount of plants that will be grown. This will determine the amount of land needed and the equipment that will be required. It is also important to consider the water source and irrigation system that will be used, as well as any necessary permits or licenses that may be required.

Additionally, it is important to plan for potential pests and diseases that may affect the cannabis plants. This can include implementing pest control measures, such as using natural predators, as well as disease prevention techniques, such as regular monitoring and proper crop rotation.

Finally, it is important to consider the security measures that will be put in place to protect the crop. This can include implementing security cameras, alarms, and fences to deter theft or vandalism.

Overall, planning for future outdoor marijuana grows requires careful consideration of a variety of factors, including location, size, water source, pests and diseases, and security measures. By taking these factors into account, growers can set themselves up for success and ensure a successful and thriving outdoor marijuana grow operation.